Geography
for Trinidad and Tobago
Forms 1, 2 & 3

Jain Cook

Workbook

Collins

HarperCollins Publishers Ltd
The News Building
1 London Bridge Street
London SE1 9GF

HarperCollins *Publishers*
Macken House,
39/40 Mayor Street Upper,
Dublin 1,
D01 C9W8
Ireland

First edition 2021

10 9 8 7 6 5 4 3 2

© HarperCollins*Publishers* Limited 2021

ISBN 978-0-00-842016-1

Collins ® is a registered trademark of HarperCollins Publishers Limited

www.collins.co.uk/caribbeanschools

A catalogue record for this book is available from the British Library.

All rights reserved. No part of this book may be reproduced, stored in a retrieval system, or transmitted in any form or by any means, electronic, mechanical, photocopying, recording or otherwise, without the prior permission in writing of the Publisher. This book is sold subject to the conditions that it shall not, by way of trade or otherwise, be lent, re-sold, hired out or otherwise circulated without the Publisher's prior consent in any form of binding or cover other than that in which it is published and without a similar condition including this condition being imposed on the subsequent purchaser.

If any copyright holders have been omitted, please contact the Publisher who will make the necessary arrangements at the first opportunity.

Authors: Jain Cook and Sheridon King Coke
Series editor: Eartha Thomas-Hunte
Publisher: Dr Elaine Higgleton
In-house senior editor: Julianna Dunn
Proofreader: Helen Bleck and Joyce Littlejohn
Maps: Sarah Woods, Gordon MacGilp and Ewan Ross
Cover design: Kevin Robbins and Gordon MacGilp
Cover photo: Christine Norton Photo/SS
Illustration and typesetting: QBS Learning
Production: Lyndsey Rogers
Printed and bound in the UK using 100% Renewable Electricity at CPI Group (UK) Ltd

This book is produced from independently certified FSC™ paper to ensure responsible forest management.

For more information visit: www.harpercollins.co.uk/green

Contents

Unit 1	Exploring the World	4
Unit 2	Boundaries and Borders	8
Unit 3	Locating Places	14
Unit 4	The Caribbean Region	16
Unit 5	Building Map Skills	19
	Fieldwork 1	26
Unit 6	The Physical Environment	27
Unit 7	The Human Environment and Population	30
Unit 8	Physical Landforms/Features and Human Land-Use	34
	Fieldwork 2	37
Unit 9	Earth's Structure	39
Unit 10	Earth's Natural Disasters	41
Unit 11	Weather and Climate	46
	Fieldwork 3	50
Unit 12	Our Environment	53

1 Exploring the World

1 Look at the map of the world below, then answer the questions.

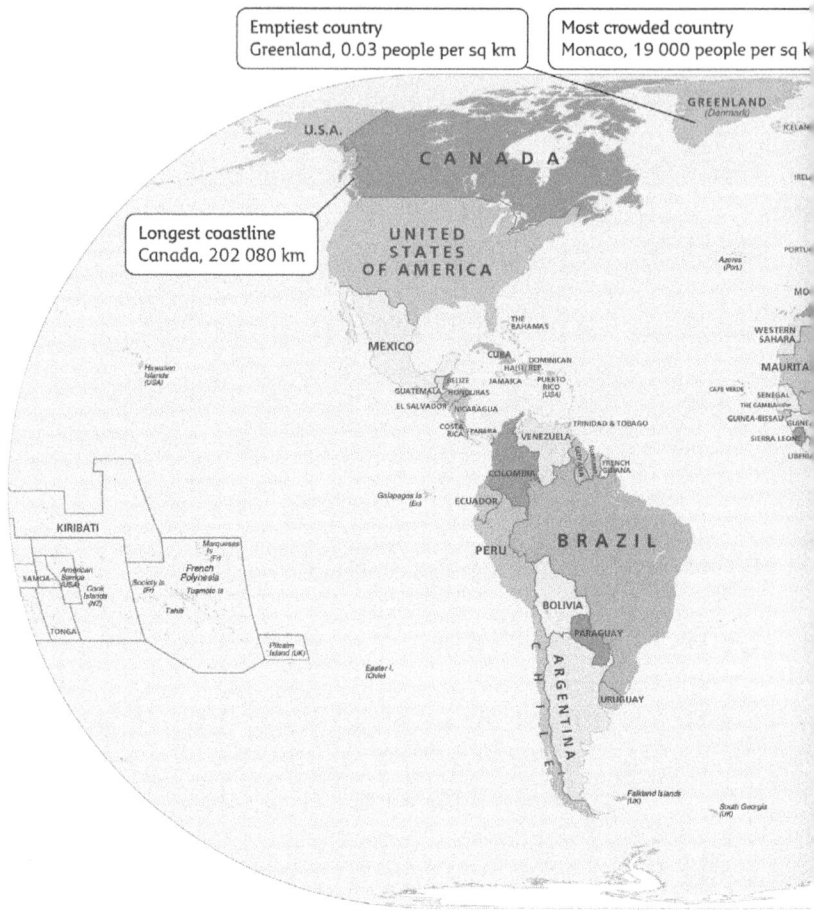

a) Mark on the map the names of the continents and oceans and at least one desert, mountain and river.

b) In which country are the most languages spoken?
_____

c) Which country is the smallest in the world?
_____

d) Which country has the least number of people per square kilometre?
_____

e) Which country has the most land boundaries?
_____

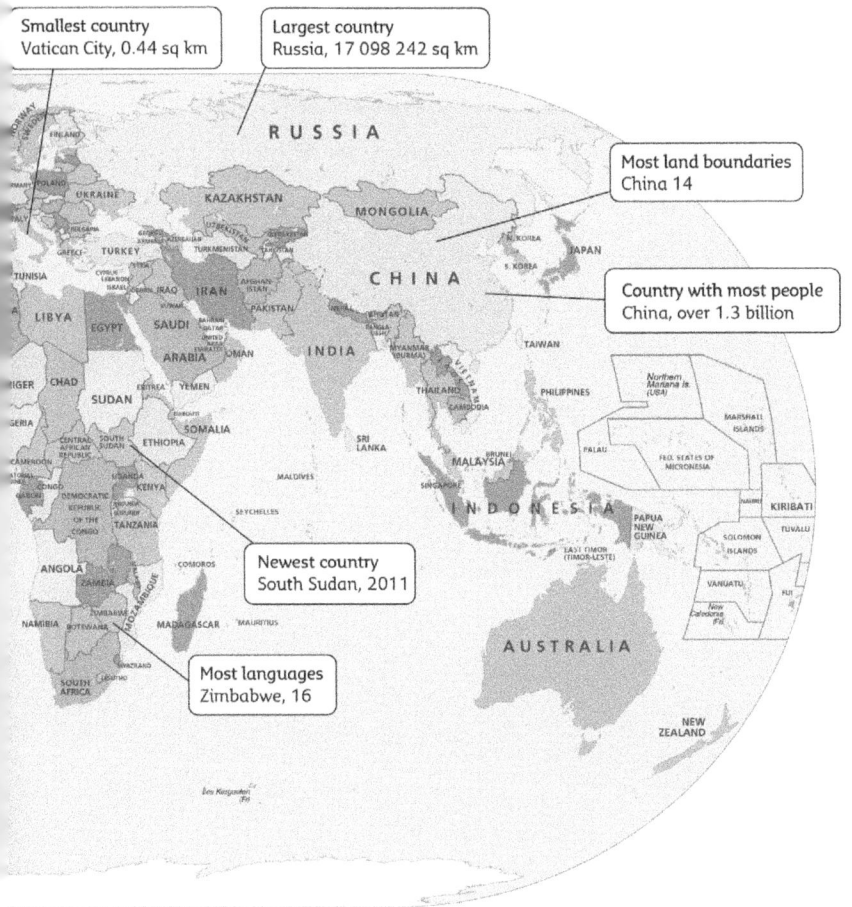

f) Name the newest country.

___

g) Which country has the most people?

___

h) Identify the largest country in the world.

___

i) Which country has the longest coastline?

___

j) Name the country that has the most people per square kilometre.

___

1 Exploring the World (cont.)

2 Complete the facts about Trinidad and Tobago. Use the internet to research any information you are not sure about.

Continent	
Population size	
Official languages	
Main religions	
Main ethnic groups	
The name of the national anthem	
The national symbol	
Currency	

3 Complete the crossword. All the words can be found in 1.1–2.3 in the Student's Book.

Across

1. a large body of land, such as a continent (8)
3. a country that is adjacent or next to another country or shares its borders (9, 5)
5. a group of people with the authority to govern a country (10)
6. a 3D map of the world (5)
7. a design or object that is a symbol of a country (6)
9. Asian country that borders Pakistan (5)
11. a geographical landform or part of the environment, such as a mountain or river (7)
12. a stream that feeds into a larger river (9)
13. the place where a river starts, usually in a mountain or hill (6)
16. an outline or dividing line; a line that encloses a given area (8)

Down

2. the idea of a nation as a whole; having a shared culture, language or any other shared traits contained in its history (8, 8)
4. one of the large, continuous expanses of land on Earth, such as South America (9)
8. a picture that shows an area of the Earth as seen from above, showing either physical or political features (3)
10. a grouping of people with common national or cultural traditions (9)
14. an examination and record of an area of land, in order to create an accurate plan or description (6)
15. a country surrounded by other countries (10)
16. edge of a country or a line separating two countries or regions (6)

2 Boundaries and Borders

1 Read carefully 2.3 in the Student's Book. Circle the correct answers below.

a) Why do countries have borders?

   i)   to separate them from other countries

   ii) to share mountain ranges

   iii) to establish territorial waters

   iv) to emphasise previous colonial rule

b) What sort of border is a mountain range?

   i)   a man-made border

   ii) a political boundary

   iii) a straight-line border

   iv) a natural boundary

c) What does a straight-line border usually indicate?

   i)   a disputed border

   ii) a neighbour state

   iii) a colonial past

   iv) a landlocked country

d) Why do border disputes arise?

   i)   people have not signed treaties

   ii) some landlocked countries need a coastline

   iii) neighbour states go to war

   iv) countries argue about who should own certain land

e) Lake Chad is the border for how many countries?

   i)   two   ii) three

   iii) four   iv) five

2 Look back at the world map on pages 18–19 of the Student's Book. See how many countries you can find for each of the letters A–Z. Add the language that country speaks. Use the internet to research any information you are not sure about.
(Clue: some letters may not have countries for them.)

	COUNTRY	LANGUAGE		COUNTRY	LANGUAGE
A			N		
B			O		
C			P		
D			Q		
E			R		
F			S		
G			T		
H			U		
I			V		
J			W		
K			X		
L			Y		
M			Z		

2 Boundaries and Borders (cont.)

3 Look at the map of Trinidad and Tobago below. Label the 11 regions, three boroughs and two cities.

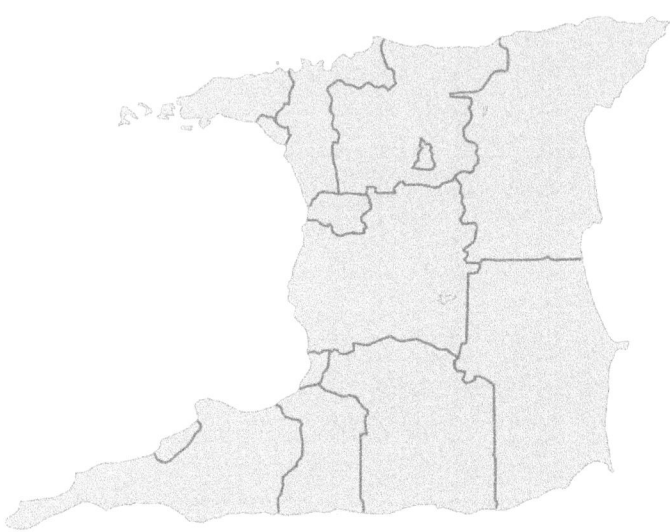

4 Now look at this map of Trinidad. Label the seven parishes of Trinidad.

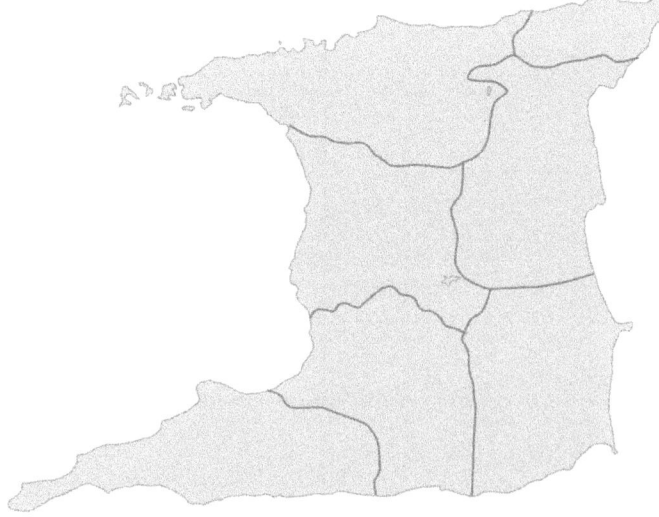

5 Read 2.3–2.5 in the Student's Book. Match the words to their definitions.

a) an elected group that governs a town

b) a smaller division within a county, usually used for voting purposes

c) a territorial division within a country, forming the main unit for local administration

d) the act of determining the boundaries or limits of an area

e) a town or district in a city that is responsible for its own schools, libraries etc.

i) borough

ii) town council

iii) demarcation

iv) ward

v) county

6 Write a letter to a pen friend overseas describing the features of Trinidad and Tobago. Include as much information as you can about the system of government, the capital city, the national anthem, flower, flag and emblem, as well as the physical features such as wetlands, rivers, mountains and coastline. Write 250 words. Use an extra sheet if necessary.

_____
_____
_____
_____
_____
_____
_____
_____
_____
_____
_____
_____
_____
_____

2 Boundaries and Borders (cont.)

7 Draw a map and invent a flag for a new country. Write a paragraph about its language(s), religion(s), ethnicity and currency.

8 Use the maps to answer the questions.

a) What type of border does Paraguay have?
_____

b) What type of border does Chile have?
_____

c) How many countries does Bolivia share its borders with? Name them.
_____
_____

d) What feature is on the Peru/Bolivia border?
_____

e) What is the nearest mainland country to Trinidad and Tobago?
_____

f) Which country shares a border with Guyana and French Guiana?
_____

g) Which countries have one of their borders demarcated by a river?
_____

3 Locating Places

1 **Look at the diagram of the Earth and label on it the following:**

The Equator   The Arctic Circle

The Tropic of Cancer   The North Pole

The Tropic of Capricorn   The South Pole

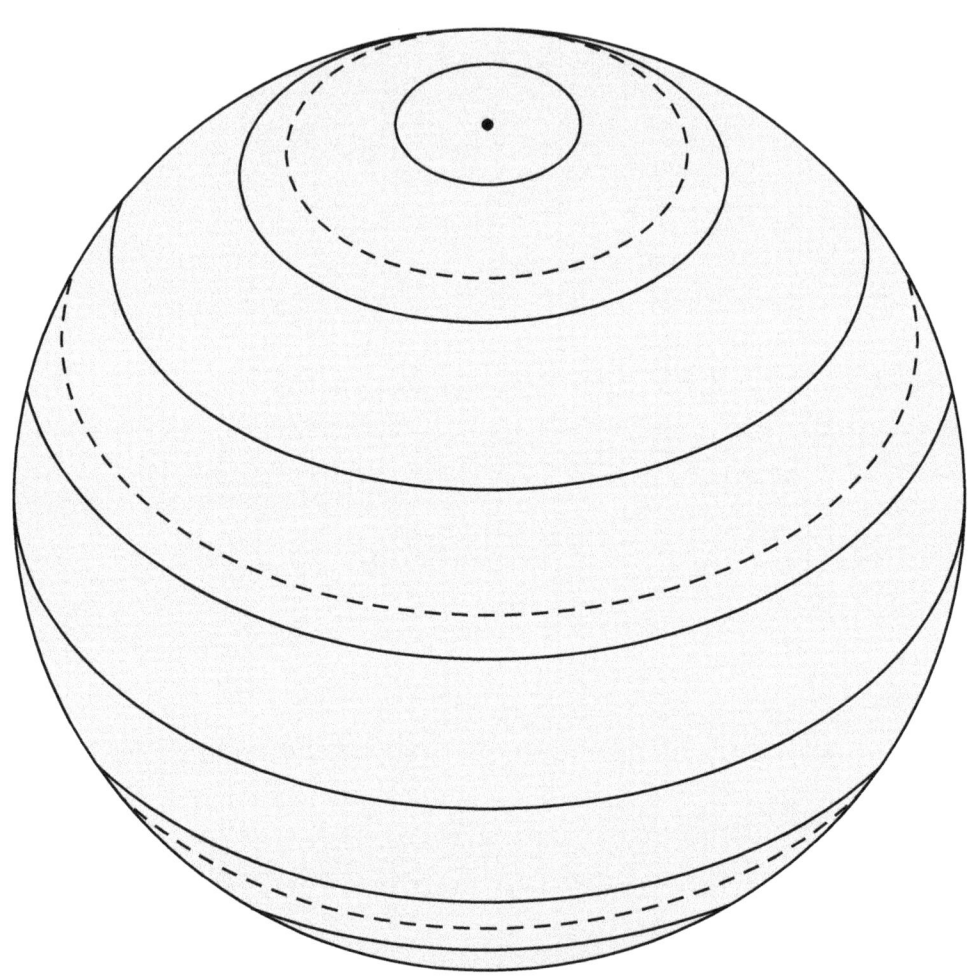

2 Use the words in the box to complete the blanks in the text.

| circumference | grid | longitude | geography |
| astronomer | west | latitude | maps |

For thousands of years, people have used the skies to help with navigation.

The Phoenicians were part of an ancient civilisation in the Middle East. They were known for exploring the world by boat. As long ago as 600 BC, the Phoenicians used the sun and stars to work out their **(1)** _____. The Polynesians also used the movement of the stars to work this out.

The ancient Greeks started using **(2)** _____ lines to show latitude and **(3)** _____. This was a suggestion by Greek astronomer Hipparchus around 300 BC. Hipparchus also found a way to locate places on Earth by observing the positions of the sun, moon and stars.

Around 225 BC, Eratosthenes, a Greek mathematician and **(4)** _____, measured the circumference of the Earth (the distance around the Earth) by calculating the distance between Alexandria in northern Egypt and Syene in southern Egypt. Once he worked this out, he was able to work out the **(5)** _____ of the Earth.

This discovery helped the Greeks draw **(6)** _____ as they were able to find their latitude easily using trigonometry and the positions of the sun, moon and stars.

Ancient scholars also made many mistakes in their ideas and writings about **(7)** _____. The Roman scholar Ptolemy believed that the circumference of the Earth was shorter than it actually is.

As a result, Christopher Columbus made the mistake of believing he could reach Asia by sailing **(8)** _____ from Europe.

4 The Caribbean Region

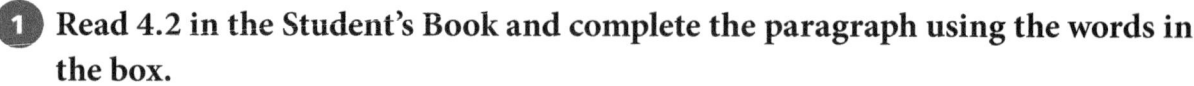

1 Read 4.2 in the Student's Book and complete the paragraph using the words in the box.

| reefs | mainland | Leeward | Antilles |
| cays | Caribbean Sea | Windward | |

The Caribbean is made up of many islands, (1) _____, and (2) _____, as well as some of the (3) _____ countries that border on the Caribbean Sea. Geographers divide the islands of the (4) _____ into two clusters: the Greater and Lesser Antilles.

The islands of the Lesser Antilles are divided into:

- (5) _____ Islands in the south
- (6) _____ Islands in the north
- Leeward (7) _____ in the west.

2 Complete the crossword of geographical terms.

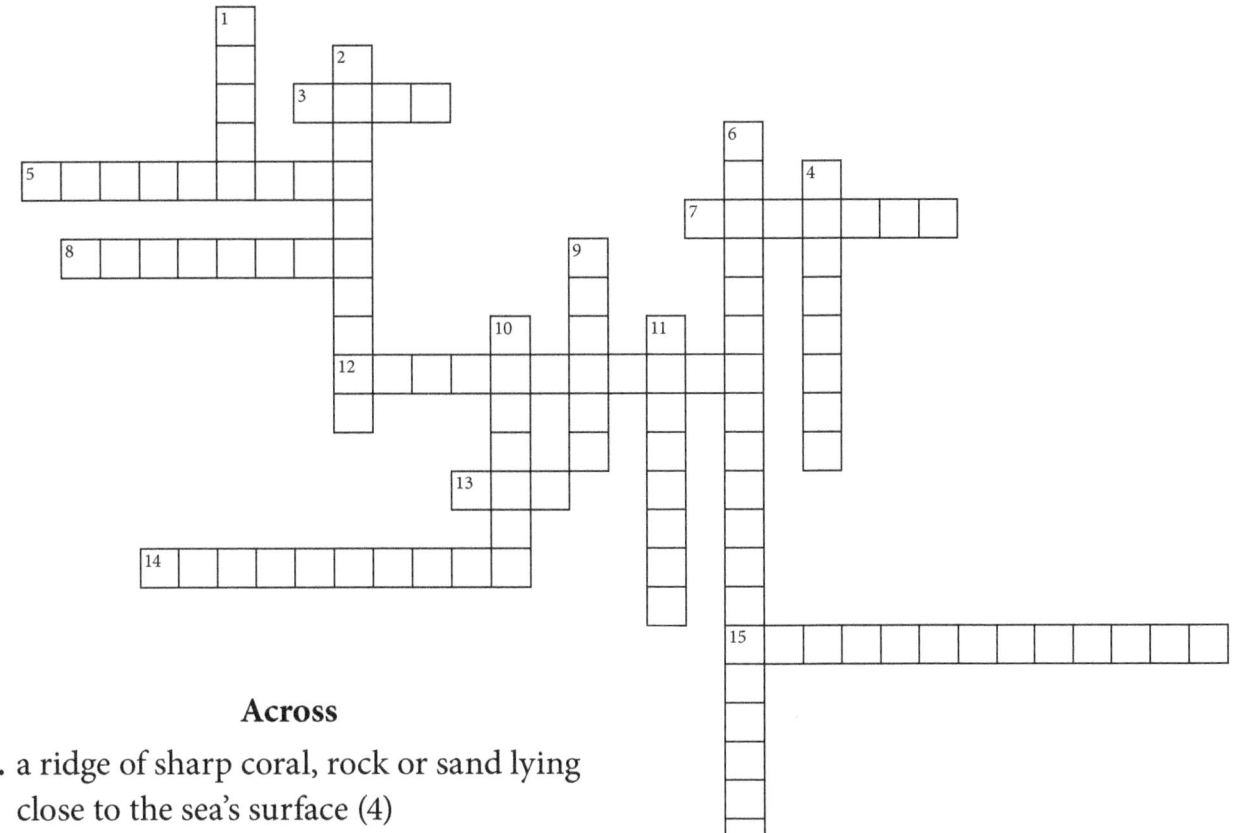

Across

3. a ridge of sharp coral, rock or sand lying close to the sea's surface (4)
5. an imaginary geographic line that runs vertically from north to south parallel to the Prime Meridian (9)
7. hemisphere where the Caribbean is situated (7)
8. a continuous stretch of land that makes up the main part of a country, as opposed to offshore islands (8)
12. working together (11)
13. a low island composed mainly of coral or sand (3)
14. joined together or working cooperatively as a single unit (10)
15. a process of making the world more connected, with goods, services and people moving and communicating easily and quickly between all parts of the world (13)

Down

1. a small island (5)
2. a country or province controlled by an outside nation (10)
4. an imaginary geographic line that runs horizontally across the globe from west to east, parallel to the Equator (8)
6. the joining or working together of countries that are geographically near each other in order to make them economically and politically more powerful (8, 11)
9. a piece of land surrounded on all sides by the sea (6)
10. towards the side sheltered by the wind (7)
11. into the wind; on the side facing the wind (8)

4 The Caribbean Region (cont.)

3 Read carefully 4.3 and 4.4 in the Student's Book. Look at the words in the box to do with the Caribbean region and integration. Then write them under the correct heading in the table.

free trade	diversifying trade	hurricanes
import and export of goods	liberalising trade	cricket
geography	languages	earthquakes
increasing trade	limited resources	storms
import and export of services	culture	
ensuring fair trade	shared history	
import and export of labour		

SIMILARITIES	HAZARDS	COOPERATION	CARICOM

5 Building Map Skills

1 Read 5.1 in the Student's Book. Unscramble each of the clue words. Copy the letters in the numbered cells into the bottom row of cells with the same numbers to find the words linked with maps.

a) HICLASPYPAM

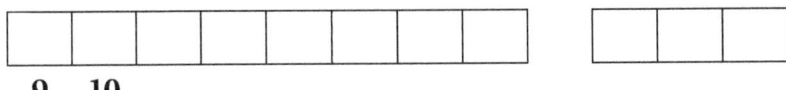

b) RODAAMP

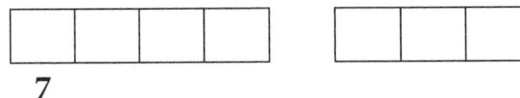

c) REEMSOASPRUC

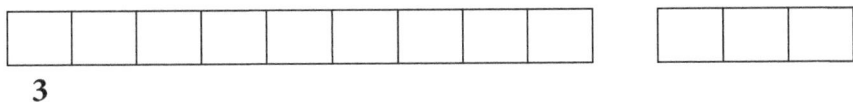

d) LAIOCILPPAMT

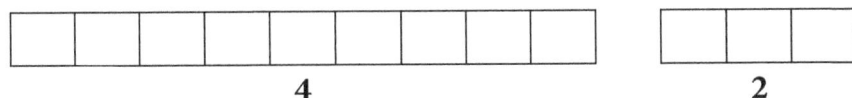

e) LYBMOS

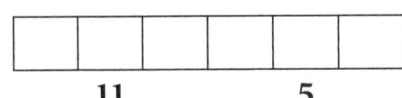

f) WROAR

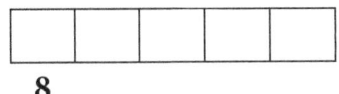

g) DEELNG

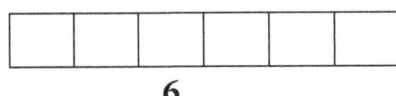

h) LACSE

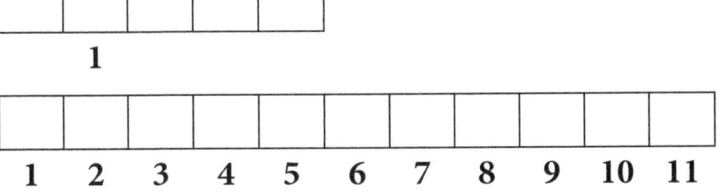

19

5. Building Map Skills (cont.)

2 Match the words about maps with their definitions.

a) bird's-eye-view _____   i) a view from directly above the ground

b) observation _____   ii) someone who plans and draws maps

c) statement scale _____   iii) a scale expressed as a number, ratio or fraction

d) to scale _____   iv) the action or process of closely and carefully looking at or monitoring something or someone

e) cartographer _____   v) the relationship between distances on a map and distances in real life

f) ratio scale _____   vi) a scale showing a line distance and what that distance represents in real life

g) linear scale _____   vii) uniformly reduced or enlarged; showing the relationships in proportion to real life

h) map scale _____   viii) scale expressed as 1 cm = 1 000 km

3 Look carefully at the maps. Match the correct scale to each map.

a) Map _____   Scale 1:50 000 000
                   0   500 1000 1500 2000km

b) Map _____   Scale 1:3 000 000
                   0   25 50 75 100 125 150km

c) Map _____   Scale 1:11 000 000
                   0   100 200 300 400 500km

Map 1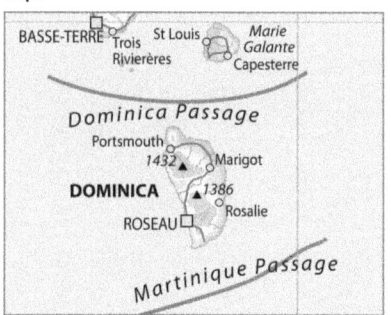

Map 2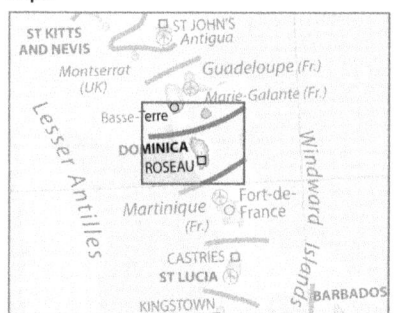

Map 3

4 Look at the map and answer the questions. (Hint: look at 5.3 in the Student's Book about ratio scale.)

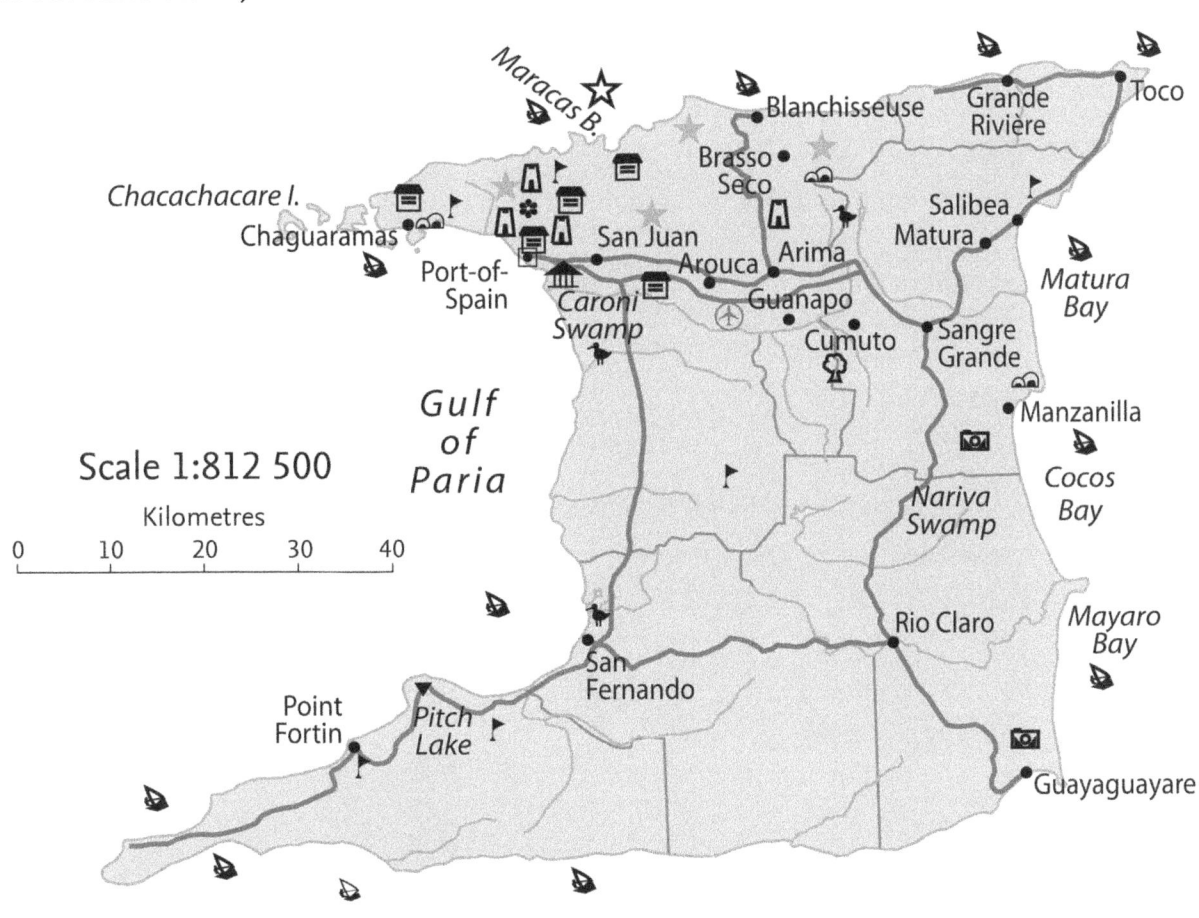

a) What is the map scale?

_____

_____

b) How far is the Pitch Lake from Cocos Bay?

_____

c) If you drove from Toco to Guayaguayare, how far would you have driven?

_____

d) Imagine you have a boat and are going to sail from Port-of-Spain to Point Fortin. How many kilometres would you have to sail?

_____

e) What is the distance between San Fernando and Rio Claro?

_____

5 Building Map Skills (cont.)

5 Follow the instructions to complete the activity.

a) Write the names of the four main cardinal points on this map.

b) Write the initials of the cardinal points and the names of the four intermediate points on this map.

c) Write the initials of the cardinal and intermediate points and the names of the secondary intermediate points on this map.

6 Read 5.6 in the Student's Book and complete the sentences.

a) Which direction is between south-south-east and south-south-west? _____

b) North-north-east is between _____ and _____.

c) South-east is between _____ and _____.

d) North-east is halfway between north and _____.

e) What direction is between south-east and south-west? _____

f) Which cardinal point comes between east-north-east and east-south-east? _____

7 Look at the map and answer the questions (Hint: See 5.3 in the Student's Book about using grid lines.)

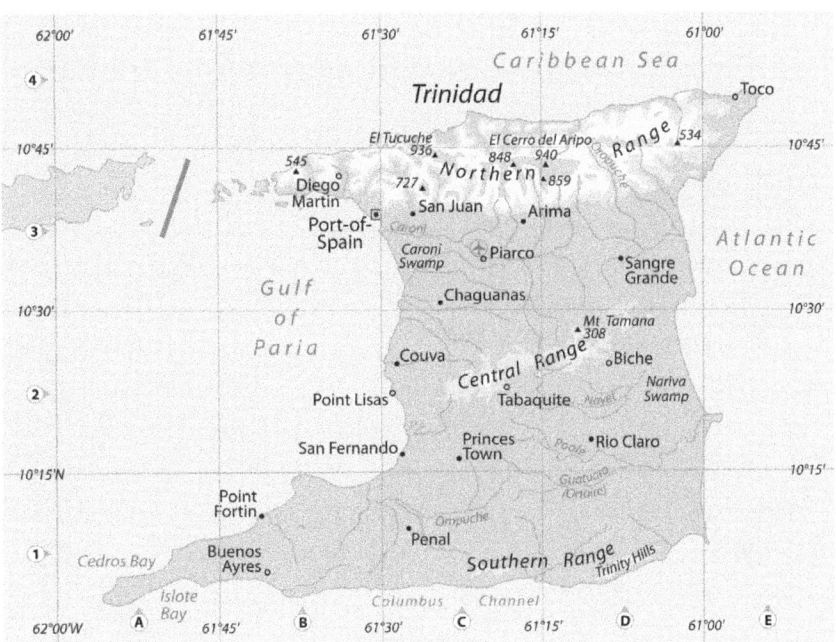

a) What gulf can be found in B3? _____

b) Name the hills found in D1. _____

c) Which swamp can be found in C3? _____

d) Which two towns are in C2? _____

e) Which bays are found in A1? _____

8 Using the map above, write down the co-ordinates for the following places.

a) Rio Claro

b) Port-of-Spain

c) The International Airport

d) Toco

e) Mt Tamana

5 Building Map Skills (cont.)

Study the map of Nanny Town and answer the questions which follow.

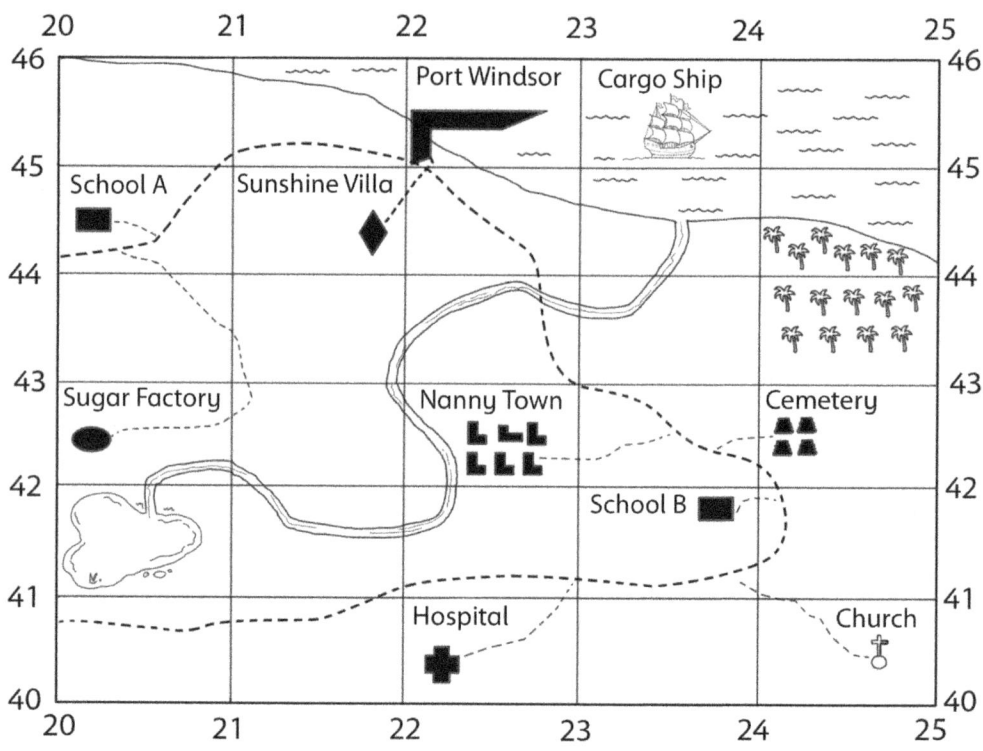

Key

(pond symbol)	Pond
(river symbol)	River
(palm trees symbol)	Coconut Plantation
(dashed line)	Road

9 Give the four-figure grid reference for the following features on the map.

  a) Hospital _____

  b) Church _____

  c) School A _____

  d) Cemetery _____

  e) Port _____

10 Which features can be found in the four-figure grid square?

  f) 2242 _____

  g) 2144 _____

  h) 2341 _____

  i) 2345 _____

  j) 2041 _____

Fieldwork 1 – Map Skills and Interpretation

1 Use the space in the box provided to draw a simple sketch map (plan view) of your classroom. Include the following in the map: *key, scale, title* and *compass direction*. The map already has a border. Keep the map as simple as possible by including only the most outstanding features in the classroom (*desk, chairs, window, chalkboard, etc.*). Ensure that you use a pencil and try to work as neatly as possible.

2 Answer the following questions based on the *Treasure Hunt* **group activity on page 72 of the Student's Book.**

   a) State the main challenge that you faced during the activity.

   _____

   _____

   _____

   b) What was the feedback from your classmates regarding the map that was drawn by your group?

   _____

   _____

   _____

   c) How would you adjust your activity to make it more effective? Explain.

   _____

   _____

   _____

6 The Physical Environment

1. Read 6.1 and 6.2 in the Student's Book. Then look at the map below of Trinidad and Tobago. Use the items in the box to label the map.

Nariva Plain	Pigeon Peak	Caroni Plain
Southern Range	Coffee River	Mount Tamana
Trinity Hills	Salybia Bay	Brigand Hill
Ortoire River	Mayaro Beach	Main Ridge

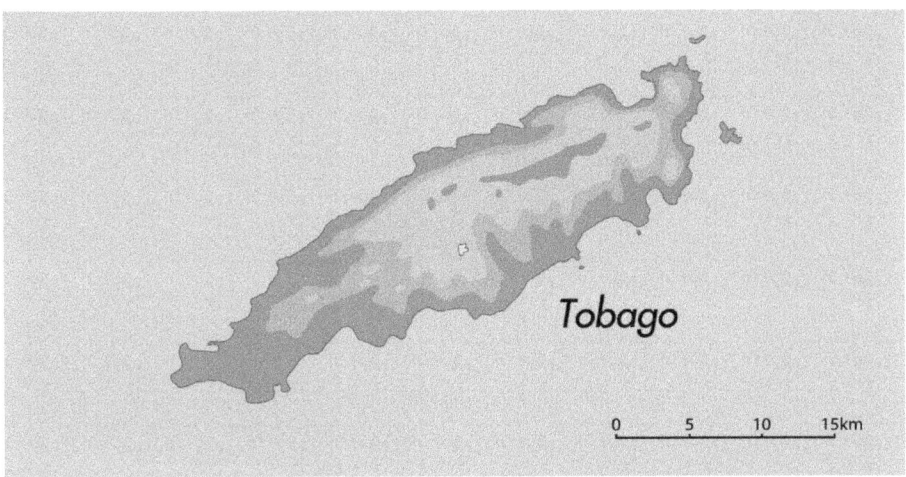

6 The Physical Environment (cont.)

2 Complete the crossword. All the words are key vocabulary words from 6.1–6.4 in the Student's Book.

28

Across

3. Work that involves going outside the classroom or laboratory to study something in a real environment (9)
5. Natural structures like very big hills that are much higher than the usual level of land around them (9)
8. A large area of water that flows towards the sea (5)
10. To put notes in a piece of writing in order to explain parts of it (8)
11. The start of a river (5)
12. To find information that is stored on a computer in order to use it again (8)
15. The process where you observe people or places (or landform features) in real locations and situations (5, 11)
16. A narrow piece of land that sticks out into the sea (8)
17. To say where you found information (4)
18. A computer program used for looking for information on the Internet (6, 6)

Down

1. A large area of land that continues further out into the sea than the land it is part of (4)
2. The area of land by the sea (9)
4. To watch or study something with care and attention in order to discover something (7)
6. The information that needs to be entered on the Internet to find a particular company (7, 7)
7. An area of sand or small stones beside the sea or a lake (5)
9. An area of the coast where the land curves inwards (3)
13. A large flat area of land (5)
14. The place where a river is widest and joins the sea (5)
15. A simple drawing that is used to highlight specific features or landforms or to show the general landscape (5, 6)

7 The Human Environment and Population

1 Unscramble each word and then match them with the correct definition below. All the words are key vocabulary words from 6.7 and 7.1 in the Student's Book.

a) gholidrolacy clyec _____

b) ryatrite triunidess _____

c) skoctevil gnamfir _____

d) mahnu verintemonn _____

e) luosindtar aetset _____

f) tentlmetes _____

g) myarirp dressuniti _____

h) clatipa tyic _____

i) almthe _____

j) bralea nimfrag _____

i) Service organisations that involve selling products

ii) Growing of crops such as sugar

iii) A continuous process that moves water between the atmosphere, the surface of the Earth and the spaces under the Earth's surface

iv) An area where people work and live

v) Involving extracting natural resources, such as mining

vi) Where the government is located

vii) The rearing of animals, such as sheep

viii) An area of land where companies have their buildings

ix) The smallest type of settlement

x) Place where people settle down and live

2 **Fill in the gaps in the map about communications on page 95 of the Student's Book with the following information:**

a) Both capital cities
b) Two important towns
c) The main highway
d) Two main roads
e) Both airports
f) Two ports
g) Caves
h) A waterfall
i) A bird sanctuary
j) A forest reserve
k) An historic site
l) A bathing beach
m) A ferry port

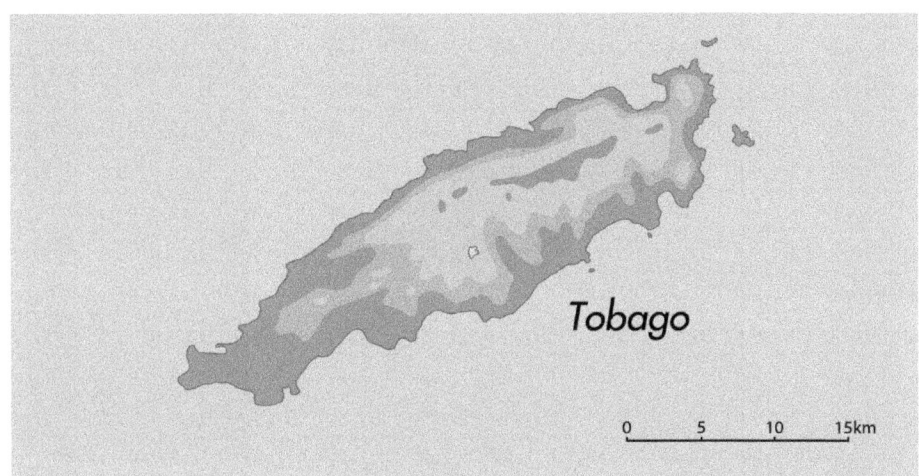

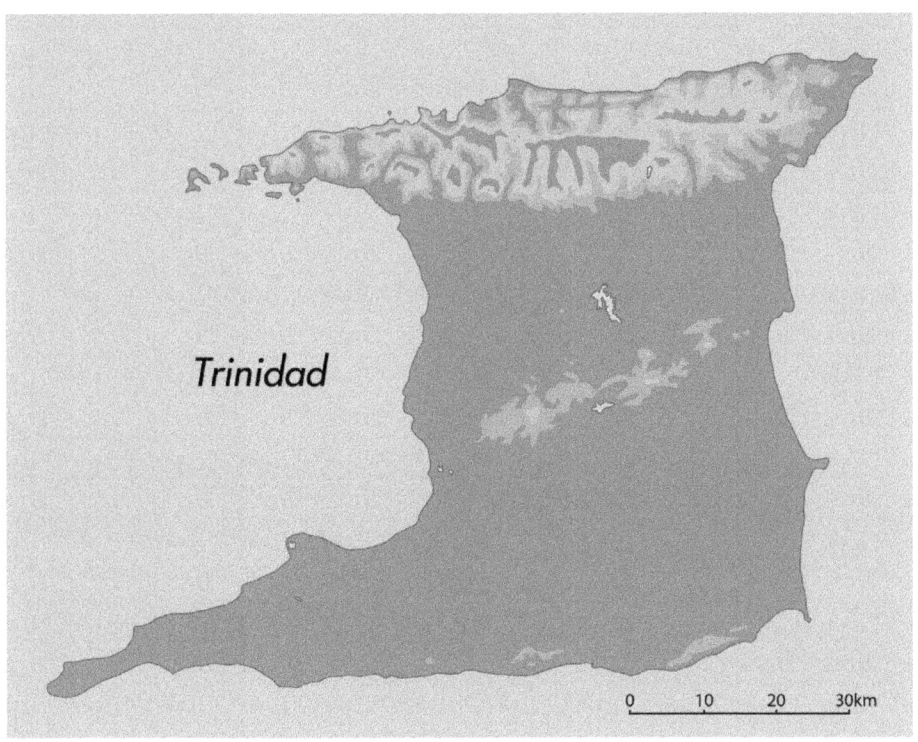

7 The Human Environment and Population (cont.)

3 Label the diagrams with the names of the types of settlement they show.

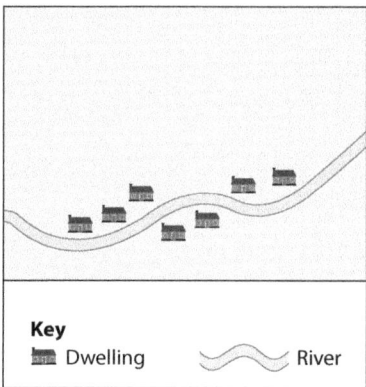 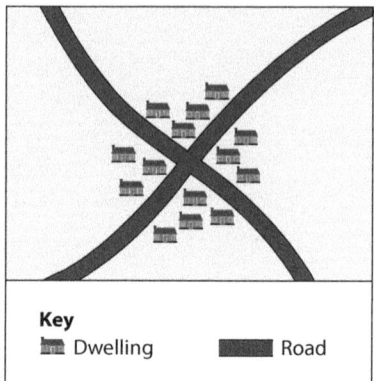 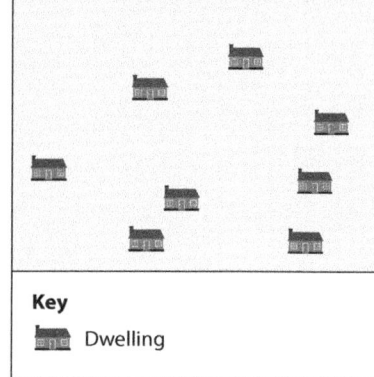

_____   _____   _____

4 Write about the area where you live. Write 200–250 words.

Things to include:
- transport facilities and nearby roads, ports or airport
- the nearest large city
- nearby recreational facilities
- the settlement pattern
- the population of where you live
- the population density of the area
- the population distribution.

5 Match each word from 7.2 and 7.3 to the correct definition.

a) Linear settlement   i) Houses and buildings grouped close together

b) Nucleated settlement   ii) Spread apart over a large area

c) Scattered settlement   iii) Where houses and buildings are built in lines

d) Settlement pattern   iv) Shape of the hamlet, village, town or city

6 Look at 7.4–7.6 in the Student's Book. Circle True or False for each of the statements below.

a) Approximately 7 billion people live in India and China. True False

b) The population of Trinidad and Tobago is around 1.4 million. True False

c) Urban areas have moderate and high population densities. True False

d) A chloropeth map shows high density areas in lighter colours. True False

e) One dot on a dot map usually represents 1 000 people. True False

f) Mountainous areas tend to have higher populations than flatter areas. True False

g) Savannah areas are generally less populated. True False

h) In Trinidad and Tobago, the coastal areas are more densely populated. True False

i) A major drawback with making dot maps is that they take a long time to make. True False

j) Rural areas have more services than urban areas, and more people. True False

7 Use the words in the box to complete the blank in each sentence.

| rural area | population | census |
| population density | population distribution | urban area |

a) _____ is not evenly spread throughout the world, with some areas very densely populated.

b) They live in a very _____, with their nearest neighbour a mile away.

c) In any capital city, the _____ is very high in each square kilometre.

d) A _____ is a way of collecting population information of a country.

e) The _____ of Trinidad is higher than that of Tobago.

f) Some people prefer to live in an _____, where it's easier to find a job and there's lots of entertainment.

8 Physical Landforms/Features and Human Land-Use

1 Match the words in the box to the correct heading in each circle. The words are from 7.6–8.1 in the Student's Book.

dams	fertile farmland	not too wet or dry
highways	hilly regions	access to services
canyons	not too hot or cold	cliffs
moderate population	larger settlements	wash
less rainfall	monuments	rivers
valleys	fish	bridges

Agriculture and soil
_____
_____
_____

Fresh water
_____
_____
_____

Climate
_____
_____
_____

Communications
_____
_____
_____

Physical features
_____
_____
_____

Man-made features
_____
_____
_____

34

2 Read 8.1–8.3. Then, match each sentence beginning (a–k) with the correct sentence ending (i–xi).

a) Areas covered in thick, dense forest _____

b) Physical features are any naturally forming features on the Earth's surface, such as _____

c) Fewer trees leads to more carbon dioxide, _____

d) Deforestation leads to other problems, such as _____

e) The land beside the sea was once covered in mangroves _____

f) Low-lying coastlines have far more opportunities to farm or build _____

g) Farming is very difficult _____

h) Man-made features are those that have been created entirely by humans, such as _____

i) The coastline beside the Port of Spain coast _____

j) When the trees are gone, _____

k) It is not possible to use this land [that floods] _____

i) so temperatures increase.

ii) provides a natural harbour for maritime traffic.

iii) skyscrapers, monuments, cathedrals, highways, bridges, ports, canals and dams.

iv) than the higher ones.

v) for any form of building.

vi) are not suitable for building on, unless the forest is cleared first.

vii) the risk of flooding increases.

viii) and completely unsuitable to build on.

ix) in mountainous and hilly areas.

x) increased levels of carbon dioxide, loss of habitats for animals and an increased risk of flooding.

xi) mountains, lakes, rivers, valleys, beaches, canyons, caves and cliffs.

8 Physical Landforms/Features and Human Land-Use (cont.)

3 Complete the words using the clues given. They are from the key vocabulary in 7.7–8.4.

a) The temporary or permanent movement of people from one place to another

_ _ _ r _ _ _ _ n

b) Something that can be continued without destroying the resources that make it possible

_ u _ _ _ _ n _ _ _ _ i _ _

c) The outside edge of an area of land

_ _ _ _ _ m _ _ _ r

d) Features created by humans such as buildings, roads, ports and airports

_ a _ - _ _ _ _ e _ _ _ _ u _ _ _

e) The level of comfort and wealth that a person or family may have

_ _ _ _ _ d _ _ _ _ _ _ _ _ i _ i _ _

f) To create more land by stopping the sea from covering it

_ _ _ _ _ a _ _

g) The process by which water is carried away from an area

_ _ _ _ i _ a _ _

h) Having a clean supply of water and good sewage system

_ _ n _ _ _ _ i _ _

i) Maintaining an original state

_ _ _ s _ _ _ a _ _ _ _

j) To make something, for example a river, change direction

_ _ _ e _ _

Fieldwork 2 – Primary Data

1. Working in groups, conduct a simple census of the number of males and females in different classes at your school. Each group would choose (one) class for their study. Your task would include the following:

   - collecting the primary data from each class.
   - creating a table showing the number of males and females in Microsoft Excel
   - presenting the data using a suitable type of graph/chart.

2. State two important findings of the census conducted in exercise 1.

   _____

   _____

   _____

3. List at least two challenges that you encountered while conducting this field study.

   _____

   _____

   _____

4. Draw a bar chart in the space provided to represent the following data in the table displayed. Ensure that your graph has the following: both axes labelled, a suitable title and appropriately shaded bars.

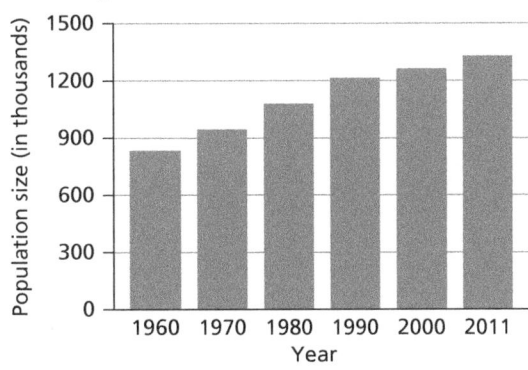

Census Population Sizes for Trinidad and Tobago: 1960, 1970, 1980, 1990, 2000, 2011

Source: Central Statistical Office: Trinidad and Tobago

Fieldwork 2 – Primary Data (cont.)

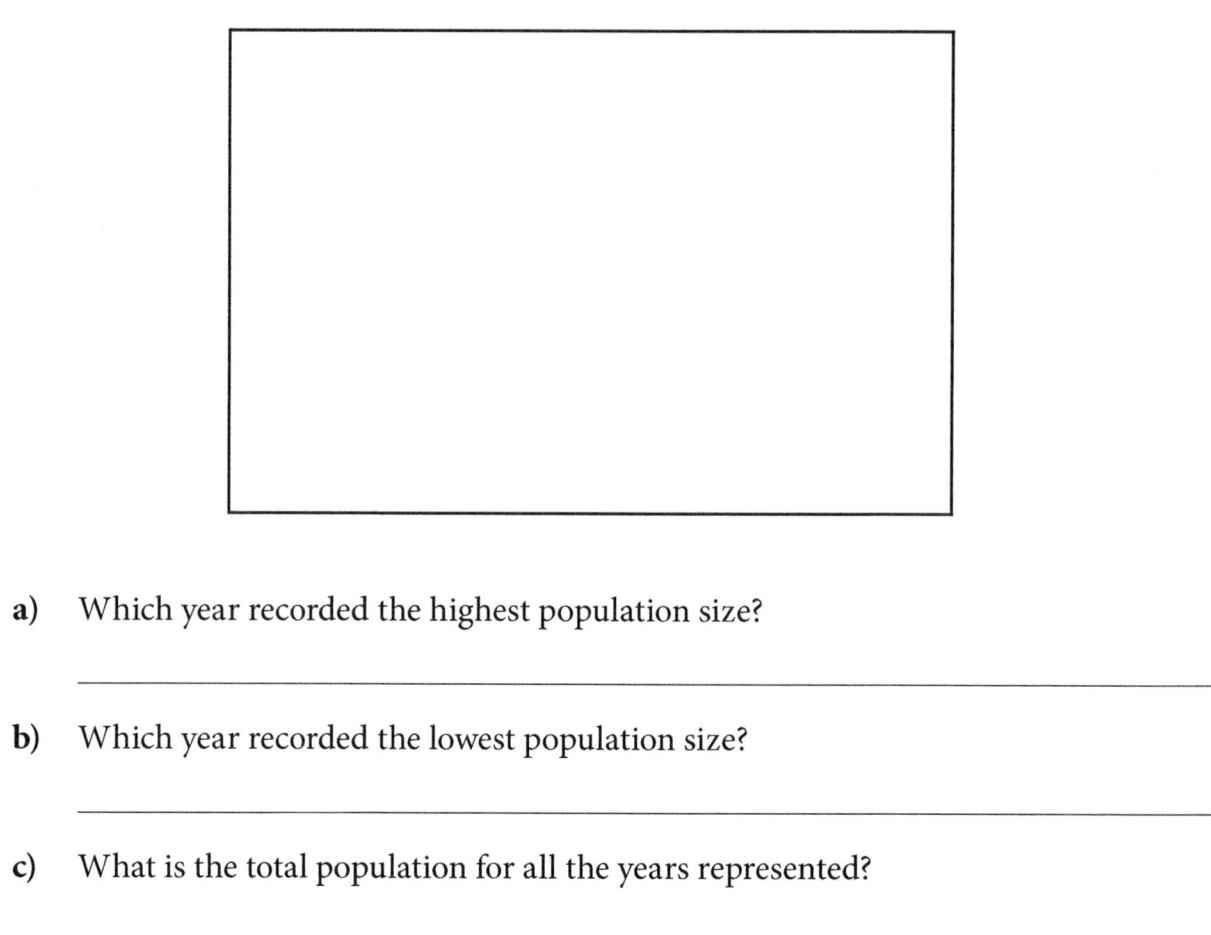

a) Which year recorded the highest population size?
_____

b) Which year recorded the lowest population size?
_____

c) What is the total population for all the years represented?
_____

9 Earth's Structure

1 Read 9.1 and 9.2 in the Student's Book. Label the diagram below and complete the answers.

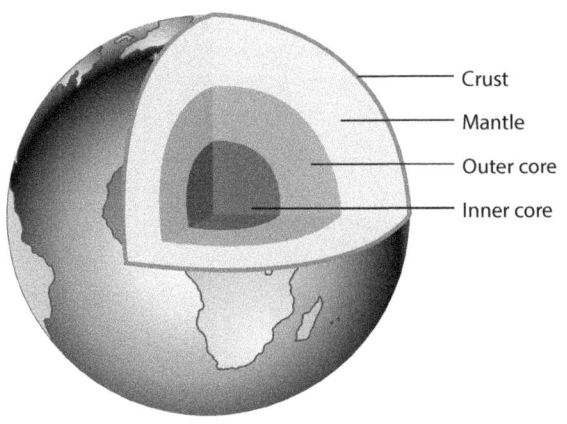

The Earth

a) _____

b) _____

c) _____

d) Name two types of crust.

_____

_____

e) Liquid rock in the mantle is known as _____.

f) The two parts of the core are **i)** _____ and **ii)** _____.

g) The Earth's crust is made up of **i)** _____ main plates and many more smaller plates. These are called **ii)** _____ plates.

h) The plates move because they are being pushed and pulled in different directions by _____ inside the mantle.

i) Where two tectonic plates meet it is called a **i)** _____ or **ii)** _____.

j) _____ or transform plate margins is when two plates slide past each other.

k) _____ (or convergent) plate margins is when two plates move towards each other.

9 Earth's Structure (cont.)

2 Read 9.3 in the Student's Book and complete the answers.

a) A country where new land or volcanic islands appear is _____.

b) The _____ are a good example where fold mountains have formed.

c) _____ has both volcanoes and earthquakes due to convergent plate margins.

d) The San Andreas Fault causes earthquakes because a conservative plate margin means two plates _____ past each other.

e) Cracks in the Earth's crust are called _____.

f) _____ is the downward movement of an oceanic plate underneath a continental plate into the Earth's mantle in certain areas of the world.

g) A series of waves that form in the oceans after an earthquake or underwater volcanic eruption is known as a _____. These waves travel at speeds of more than 950 km per hour.

h) Although there are many types of rock on the Earth's surface, they fall mainly into one of three categories: **i)** _____, **ii)** _____ or **iii)** _____.

i) There are two types of igneous rock: **i)** _____ and **ii)** _____. These rocks are formed inside the Earth, beneath the surface.

10 Earth's Natural Disasters

1 Read the key vocabulary in 9.4 and 10.5 in the Student's Book. Then complete the crossword.

Across
8. Rocks that form above the surface of the Earth (9, 4)
9. Sediment glued together by crystals that form a rock (11)
11. Instruments that record the strength of seismic waves (12)
16. Formed under the seas and oceans from sediment (11, 5)
17. Sudden shaking of the Earth's crust (10)
18. Igneous rocks that form inside the Earth (9, 4)

10 Earth's Natural Disasters (cont.)

Down

1. Formed by pressure or heat from underneath the surface of the Earth (11, 5)
2. Became solid or made something solid (10)
3. The point directly above the focus on the Earth's surface (9)
4. Formed when magma from the mantle rises to the surface, then cools and hardens (7, 5)
5. The size of an earthquake (9)
6. A means of knowing the size of an earthquake (7, 5)
7. The result of the movement of tectonic plates, such as an earthquake (7, 8)
10. The point at which an earthquake happens underground (5)
12. Relating to earthquakes (7)
13. Levels of energy that travel through the Earth's layers, as the result of an earthquake (7, 5)
14. Sediment laid down or left behind on the ocean floor (9)
15. Carried or moved along into the seas and oceans (11)

2 Read 10.2–10.4 in the Student's Book. Use the words in the box below to complete the blank in each sentence.

mitigation strategy	evacuation plan	landslide
emergency procedures	Action Plan	insurance	vulnerable
aftershocks	devastating	mudslide	prone to	sway

a) Most people that can afford it take out _____ against earthquake damage.

b) All schools in Japan have an _____ for when an earthquake strikes.

c) The result of the earthquake in Haiti was _____ , with thousands made homeless.

d) Schools and businesses throughout Asia have an _____ if there is the threat of a tsunami.

e) After the main earthquake in Haiti, there were more than 50 _____ .

f) A _____ is a plan to reduce the loss of life and property by lessening the impact of disasters.

g) More than 100 tonnes of wet soil moved down the hill in the _____ that happened after the earthquake.

h) In areas that are _____ earthquakes, many people have a disaster supply kit packed and ready for use in the event of a disaster.

i) Many of the roads were blocked because of fallen rocks and debris caused by the _____ .

j) Many islands throughout the Caribbean are _____ to earthquakes and tsunamis.

k) It is essential to have informed all students or staff about the _____ we have in place.

l) Japanese architects now design buildings that will _____ when there is an earthquake.

10 Earth's Natural Disasters (cont.)

3 Read 10.4–10.6 in the Student's Book. Then answer the questions below.

a) Which items are recommended for all classrooms to have to be prepared for an emergency?

i) _____ ii) _____

iii) _____ iv) _____

v) _____ vi) _____

vii) _____ viii) _____

b) When should you not help others in the event of an emergency?
_____

c) Name the three types of volcano.

i) _____ ii) _____

iii) _____

d) Where do shield volcanoes form?
_____

e) Where do composite volcanoes form?
_____

f) How is a composite volcano structured?

i) _____

ii) _____

iii) _____

g) What did the eruption in Soufrière Hills in 2010 lead to?
_____

h) What ripped through the Belham Valley in 2009?
_____

i) What materials, mentioned on page 148 of the Student's Book, can be ejected from a volcano?

i) _____ ii) _____ iii) _____

4 Read 10.6 and 10.7 in the Student's Book. Then match the beginning of the sentences in Column A with the endings in Column B.

Column A	Column B
a) Volcanoes frequently emit carbon dioxide,	i) is a highly toxic gas.
b) Small emissions of carbon dioxide	ii) one of the deadliest effects of volcanic eruptions.
c) Hydrogen sulphide	iii) and ejected from composite volcanoes.
d) Pyroclastic flows are	iv) will not harm people.
e) Lahars are fast-moving flows of water and volcanic debris	v) which can lead to headaches and dizziness.
f) High viscosity lava is very thick	vi) that can completely bury buildings and roads.
g) Large amounts of gas and dust in the atmosphere	vii) can stop sunlight getting through to the Earth's surface.
h) If ash gets into a plane's engine,	viii) it can cause it to stall and crash.

5 Imagine that you are in charge of disaster response after an earthquake or a volcanic eruption. Identify the four activities you would complete first as a priority. Write 250 words.

Think about:
- setting up temporary hospitals
- feeding the homeless and hungry
- restoring the telephone system and electricity
- getting the TV stations running again
- rescuing everyone
- distributing drinking water
- repairing roads, ports and airports to reopen them, or opening schools.

Add any ideas of your own.

11 Weather and Climate

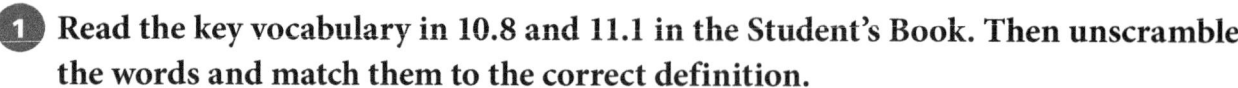

1 Read the key vocabulary in 10.8 and 11.1 in the Student's Book. Then unscramble the words and match them to the correct definition.

a) c r e p i n t a p o i t i   _____

b) o t h e r g l a m e   g y e r n e   _____

c) t h a w e r e   _____

d) p r e m a t u r e t e   _____

e) f l e t i r e   _____

f) d r a s t a u e t   _____

g) m e a t l i c   _____

h) t h i y u m d i   _____

i) i r a   e r r u p e s s   _____

j) s t e m e n l e   f o   t h a w e r e   _____

i) Soil that is rich in nutrients, so plants grow very well

ii) A pattern of weather in a place over a longer period

iii) A set of conditions in the atmosphere at a particular time and place

iv) The weight of the Earth's atmosphere on the surface

v) The measure of heat energy in the atmosphere around the Earth

vi) Rainfall

vii) When air holds the maximum amount of water vapour possible

viii) The measure of how much water vapour is in the air at any given time

ix) Water that is heated underground and produces steam and can be used to generate electricity

x) Made up of temperature, precipitation, wind, sunshine, cloudiness, humidity and air pressure

2 Read 11.2 and 11.3 in the Student's Book. Then circle the word or phrase in each group below that does not belong with the rest. Say why the item you circled does not belong. On the lines provided, write a sentence for each circled word or phrase.

a) i)   Four seasons   _____
    ii) Rainfall all year round _____
    iii) Winter months cold   _____
    iv) Humid   _____

b) i)   Horizontal belts   _____
    ii) Seasons   _____
    iii) Rainfall statistics   _____
    iv) Latitudes   _____

c) i)   Frostbite   _____
    ii) Skin cancer   _____
    iii) Malaria   _____
    iv) Heatstroke   _____

d) i)   Animal hides   _____
    ii) Range of temperatures   _____
    iii) Cotton or linen   _____
    iv) Wrap up warm   _____

e) i)   Hypothermia   _____
    ii) Pneumonia   _____
    iii) Desert regions   _____
    iv) Sunburn   _____

11 Weather and Climate (cont.)

3 Read the key vocabulary in 11.4 and 11.5 in the Student's Book. Then complete the words using the clues given.

a) A violent storm with extremely strong winds and heavy rain

_ _ _ _ i _ _ _ e

b) The centre of a storm

_ _ _

c) Information that shows temperature and rainfall statistics for a country or area over a period of time

_ _ _ _ a _ _ _ r _ _ _ s

d) A tropical storm with strong winds that moves in circles

_ _ p _ _ _ _

e) Changes into liquid and forms thick clouds

_ o _ _ e _ _ _ _ _

f) An area of extreme low pressure

_ i _ _ _ - _ i _ _ _

g) Information that is used to show rainfall figures

_ _ o _ _ _ _ _ _ _ s

h) Change into gas or steam

_ _ a _ _ _ a _ _ _ _

i) The variation of temperatures over a year

_ _ n _ _

j) A severe storm in which the wind spins in a circle

_ _ _ l _ _ e

48

4 Read 11.7–11.9 in the Student's Book. Then write the words in the box in the correct place in the diagram.

| rain begins   huge storm clouds form   storm surges   air pressure rises
early warning and detection   temperatures are higher
loss of life and property   winds reach their strongest force   there is no rain
disaster supply kits   air pressure starts to fall   winds die down
rain turns to showers   flooding   temperatures fall
huge, thick clouds form   it is sunny   hurricane shelters   cloud breaks up
torrential rain falls   winds increase |

Before a hurricane
_____
_____
_____

During a hurricane: The eye wall
_____
_____
_____

During a hurricane: The eye
_____
_____
_____

During a hurricane: The second eye wall
_____
_____
_____

After a hurricane
_____
_____
_____

The effects of hurricanes
_____
_____
_____

Preparing for a hurricane
_____
_____
_____

Fieldwork 3 – Secondary Data

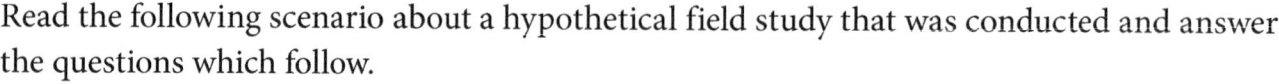

Read the following scenario about a hypothetical field study that was conducted and answer the questions which follow.

A group of 5 small farms are located in a rural area in the community of Little Village, Port of Spain. A number of vegetable crops such as tomato, cabbage, lettuce, carrot and cucumber, are grown on small plots of land (less than one hectare). These farms are faced with a number of challenges during regular farming activities. These include: natural hazards (drought and hurricane) pests and diseases, and theft (praedial larceny). A group of geography students visited the farms to investigate the various issues on September 10, 2020 between 9:00 a.m. and 12 noon. They were afterwards asked to produce a research project on the field study conducted.

1 Create a suitable aim of study for the research project.

___

2 Outline the possible methods that you would use to collect data for this study.

a) Where was the data collected?

___

b) When was the data collected?

___

c) How was the data collected?

___

3 Table 1 below is a list of the crops and the number of farms growing each crop. Use the data to draw a pie chart in the space provided.

Crops	Number of farms growing each crop
Cucumber	4
Cabbage	4
Lettuce	4
Tomato	3
Carrot	3

Table 1. Number of farms growing crops in Little Village, Port of Spain.

Fieldwork 3 – Secondary Data (cont.)

4 Figure 1 is a bar chart showing the number of farms affected by challenges in Little Village, Port of Spain. Use the data to answer the following questions.

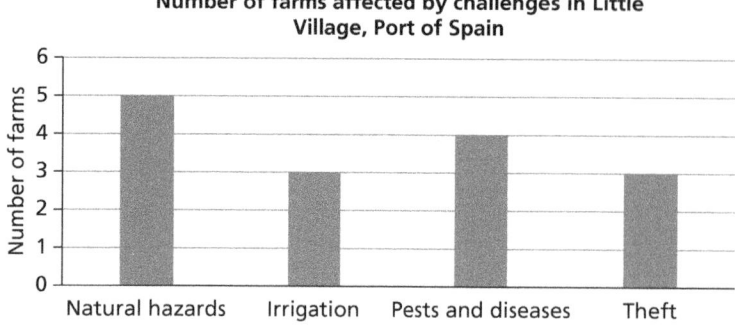

Figure 1. Number of farms affected by challenges in Little Village, Port of Spain

a) What do you think would be an appropriate label or title for the 'X' axis on the bar chart?

_____

b) Write a brief interpretation of the data presented in **Figure 1**.

_____
_____
_____
_____
_____
_____
_____

c) State at least two secondary sources of data that you could use to aid in your research of this topic.

_____
_____

12 Our Environment

1 Complete the table about natural resources that have a negative impact on the environment. Use the words in the box. All the words are from 12.6 in the Student's Book.

construction	global hydrological cycle	heats
sulphur dioxide	climate change	vehicles
soil erosion	global warming	oil spills
power	dangerous gases	furniture
flooding	planes	creates pollution

Natural resource	Used for	Impact on the environment
Coal	a) _____ and heating homes	Creates h) _____ from the thick smoke; releases i) _____
Oil	fuel to power b) _____ and c) _____; also d) _____ homes	Releases j) _____ into the atmosphere; can result in k) _____
Gas	e) _____ electricity	Emissions cause l) _____ and m) _____
Wood	f) _____ and g) _____	Affects the n) _____; threat of increased o) _____; Deforestation leads to p) _____

12 Our Environment (cont.)

2 Write the words in the box in the correct spaces below. All the words are from Unit 12 in the Student's Book. These are the five strategies that the government would like to introduce.

improve	determine	manage	establish
	reduce		

a) to _____ environmental issues better _____

b) to _____ the carbon footprint _____

c) to _____ the risk of climate change to the nation

d) to _____ effective waste management systems

e) to _____ natural resource management _____

3 Match the following to a), b) or d) in Exercise 2.

i) Use of natural resources and better waste management processes

ii) Redevelop old land and protect endangered and threatened species

iii) Use existing energy resources better

iv) Use more renewable energy

v) Use better chemical, management and climate change processes

vi) Protect coastal and marine areas and better manage biodiversity

vii) Review of environmental policy and legislation and relevant standards

viii) Reduce the rate of losing natural habitats

4 Read 12.4. Answer the following questions about natural resources and oil spills.

a) Where does oil form?

___

b) How do people get the oil out?

___

c) Where is oil usually found?

___

d) How is oil transported from country to country?

___

e) What can happen to oil that damages the environment?

___

f) Name three effects to the environment caused by oil spills.

i) ___

ii) ___

iii) ___

12 Our Environment (cont.)

5 Circle the word in each group that does not belong with the rest. They are all in Unit 12 in the Student's Book. Say why the words do not belong. On the lines provided, write a sentence using each of the words you circled.

a) i) Natural beauty _____

   ii) Culture _____

   iii) Energy sector _____

   iv) Events _____

b) i) Forests _____

   ii) Green area _____

   iii) Butterflies _____

   iv) Business and nature _____

c) i) Ports _____

   ii) Ship repairs _____

   iii) Yachting _____

   iv) Banking _____

d) i) Efficient allocation of water _____

   ii) Shipbuilding _____

   iii) Protect wetlands _____

   iv) Public education _____

e) i) Business park _____

   ii) Knowledge-based economy _____

   iii) Agro-processing _____

   iv) Export-based business _____

f) i) Water resources _____

   ii) Forest fires _____

   iii) Deforestation _____

   iv) Quarrying _____

6 Complete the crossword. All the words are key vocabulary words from Unit 12 in the Student's Book.

12 Our Environment (cont.)

Across

2. Carbon dioxide that forms naturally in the atmosphere (10, 3)
3. The continuous movement of water around the world (6, 12, 5)
7. To make air, water or land too dangerous or dirty for people to use in a safe way (7)
8. The process whereby soil is gradually removed by wind, rain or sea (4, 7)
9. Taken out (9)
10. Making use of something in order to gain as much as possible from it (12)
11. Resources that can be constantly replaced and will never be used up (9)
12. Agriculture that is based on protecting and preserving the environment (4-10)
13. When an economy is changed so that it produces goods and services for a number of different industries (8, 15)
14. Items such as coal, gas or oil (6, 5)
15. When liquid escapes from a pipeline or tanker (3, 5)
16. Items used to make something else (3, 9)
17. The slow increase in the temperature of the Earth (6, 7)

Down

1. The amount of carbon dioxide that a person, organisation or building produces and how it affects the environment (6, 9)
4. To use resources responsibly, without using them up or destroying them (11, 11)
5. Things that can never be replaced (3-9, 9)
6. Stop or catch something before it can go somewhere else (9)

7 Circle True or False for each of the statements below.

a) An objective of the National Forestry Policy is to ensure unsustainable use of forests. True   False

b) The National Forestry Policy also aims to raise public awareness of the cultural value of forests. True   False

c) The Environmental Management Authority conducts analyses of air, soil and water. True   False

d) Internal migration is when someone moves from one country to another. True   False

e) One push factor is living through a natural disaster and then moving somewhere for better living conditions – the pull factor. True   False

f) A pull factor for many people when moving from an urban area to a rural area is the possibility of finding employment. True   False

g) Pull factors encourage people to move to a new location, often for a negative reason. True   False

h) A high population density means that there are many people living per kilometre. True   False

i) Unemployment is when a person in the labour force is capable and willing to work, but has not yet found a job. True   False

j) A resort is a place where tourists can visit the natural heritage of a country. True   False

k) Ecotourism means treating the environment responsibly and carefully. True   False

8 Correct the False sentences in Exercise 7.

12 Our Environment (cont.)

9 Look at the map below and the list of tourist sites in Trinidad and Tobago. Write the number of each attraction in the correct place on the map.

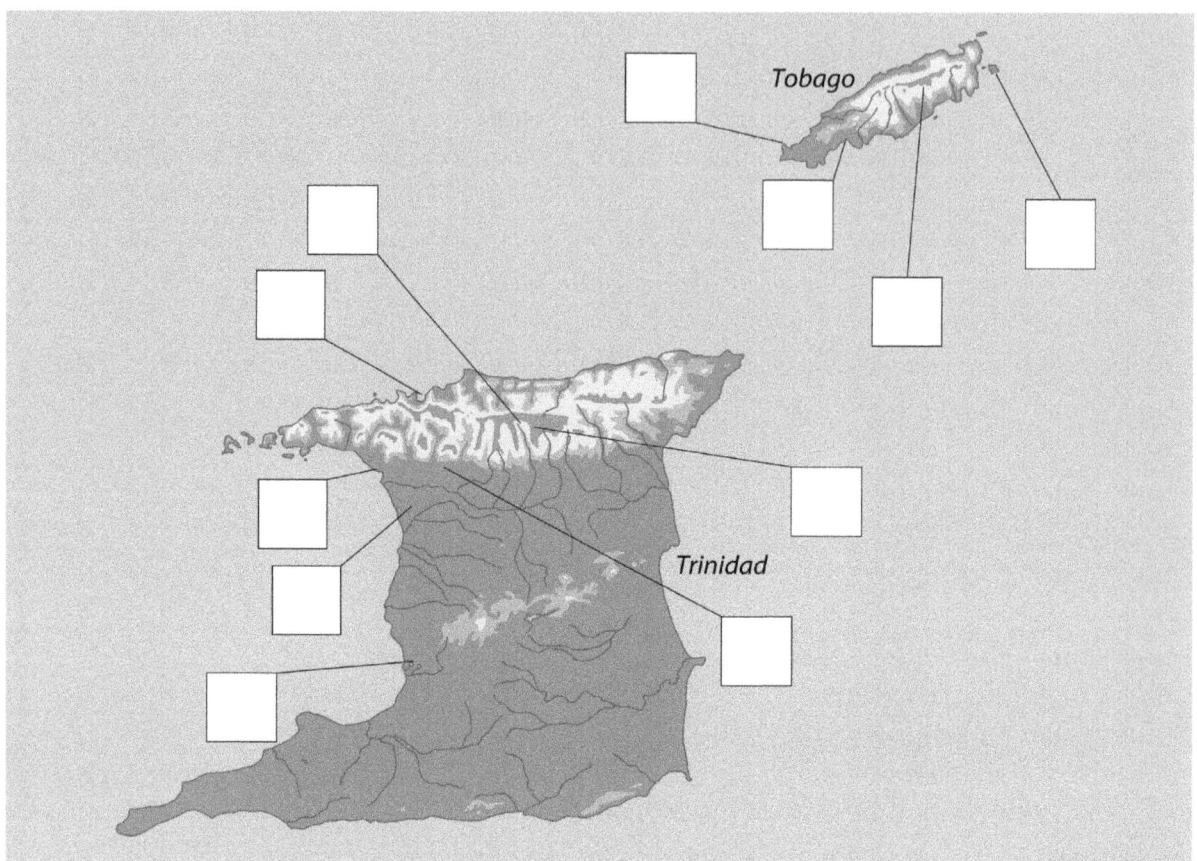

10 Write a letter to your friend who lives overseas. Tell him or her about what the government of Trinidad and Tobago is doing to diversify in the manufacturing sector.

11 Write the words from the circle below under the correct heading. All the words are from Unit 12 in the Student's Book.

Climate Flora Fauna

_____ _____ _____
_____ _____ _____
_____ _____ _____
_____ _____ _____

profitable   equatorial
tour guides   sea turtles   waterfalls
vegetation   reptiles   small businesses
entertainment facilities   bird sanctuaries   savannahs
underdeveloped   unique ecosystems   high unemployment   tropical
sandy beaches   warm and dry   shops and services
improve services   amphibians   lush rainforests
mountains   seasons   lakes

Natural sites Generation of income Job creation

_____ _____ _____
_____ _____ _____
_____ _____ _____
_____ _____ _____

12 Our Environment (cont.)

12 Find the words listed below in the word search puzzle. The words can be horizontal, diagonal or vertical and may be spelled back-to-front.

resort	push factor	rural–urban migration
conservation	climate	watershed degradation
multiplier effect	flora	pull factor
interconnectedness	ecotourism	population density
tourism	fauna	urbanisation
internal migration	unemployed	unemployment

R	S	R	E	S	O	R	T	Y	I	V	A	K	Q	G	W	D	M	D	S
N	O	I	T	A	R	G	I	M	N	A	B	R	U	L	A	R	U	R	S
K	N	C	U	H	N	Y	M	U	S	M	K	N	A	T	T	R	L	Y	E
X	M	O	H	N	P	O	Z	Y	S	Z	E	N	O	N	E	L	T	A	N
F	S	M	I	O	E	L	I	I	U	M	G	U	D	Z	R	I	I	R	D
N	K	C	H	T	I	M	R	T	P	B	R	D	F	L	S	D	P	O	E
T	R	O	S	E	A	U	P	L	A	I	A	Q	D	N	H	H	L	L	T
A	X	F	H	Q	O	R	O	L	S	S	X	M	E	A	E	E	I	F	C
D	N	E	W	T	A	Y	G	M	O	U	I	D	X	X	D	K	E	L	E
W	K	U	O	D	M	W	Y	I	L	Y	N	N	L	T	D	B	R	R	N
C	V	C	A	E	Y	T	N	H	M	O	E	U	A	A	E	P	E	O	N
S	E	U	N	F	I	Q	R	P	I	L	B	D	M	B	G	K	F	T	O
C	X	T	X	Q	V	C	Z	T	W	H	A	T	O	Y	R	W	F	C	C
L	N	Q	N	O	I	T	A	V	R	E	S	N	O	C	A	U	E	A	R
I	S	U	J	L	G	L	H	X	I	N	C	Y	R	S	D	Y	C	F	E
M	K	P	M	P	U	S	H	F	A	C	T	O	R	E	A	Q	T	L	T
A	L	F	Q	P	O	Z	W	Y	A	R	Z	G	K	T	T	F	P	L	N
T	K	X	O	Q	Y	Z	R	A	W	C	T	Q	U	L	I	N	T	U	I
E	Z	P	E	J	B	A	E	F	G	O	L	V	H	X	O	F	I	P	J
L	R	Q	D	B	A	I	E	X	Z	G	K	O	C	S	N	R	D	L	T

13 Now write the words in alphabetical order.

a) _____ b) _____

c) _____ d) _____

e) _____ f) _____

g) _____ h) _____

i) _____ j) _____

k) _____ l) _____

m) _____ n) _____

o) _____ p) _____

q) _____ r) _____

14 Complete the words using the clues given.

a) A term that refers to countries being connected

_ _ _ e _ _ _ _ _ _ _ c _ _ _ _ e _ _

b) Human activities that cause damage to bodies of water through dumping of sewage or water activities

_ _ _ _ _ r _ o _ _ _ _ i _ _

c) When the quality of the soil is reduced because of extreme weather or human activities

_ _ _ _ e _ _ a _ _ _ _ _ _ _

d) Caused by an increase of carbon dioxide in the atmosphere, especially through air transportation

_ _ _ _ _ a _ _ _ h _ _ _ _ _

e) The increased income generated by businesses associated with the tourism industry, which creates more wealth

_ u _ _ _ p _ _ _ _ _ _ _ _ e _ _

f) Dangerous or annoying levels of noise, caused by human activities

_ _ _ _ _ e _ _ l _ _ _ _ _ _ _